Robotober

BY LIZ PALMIERI-COONLEY

Robotobet

Hello,

My robot friends have come together from all over the universe to share their ABC's of science with you.

They were inspired by my niece, Honey Love, and her fascination with science.

Happy learning!

With love & gratitude,

Liz

atmosphere

the gases that surround a planet

bacteria

they are invisible and bad bacteria are the germs that make us sick

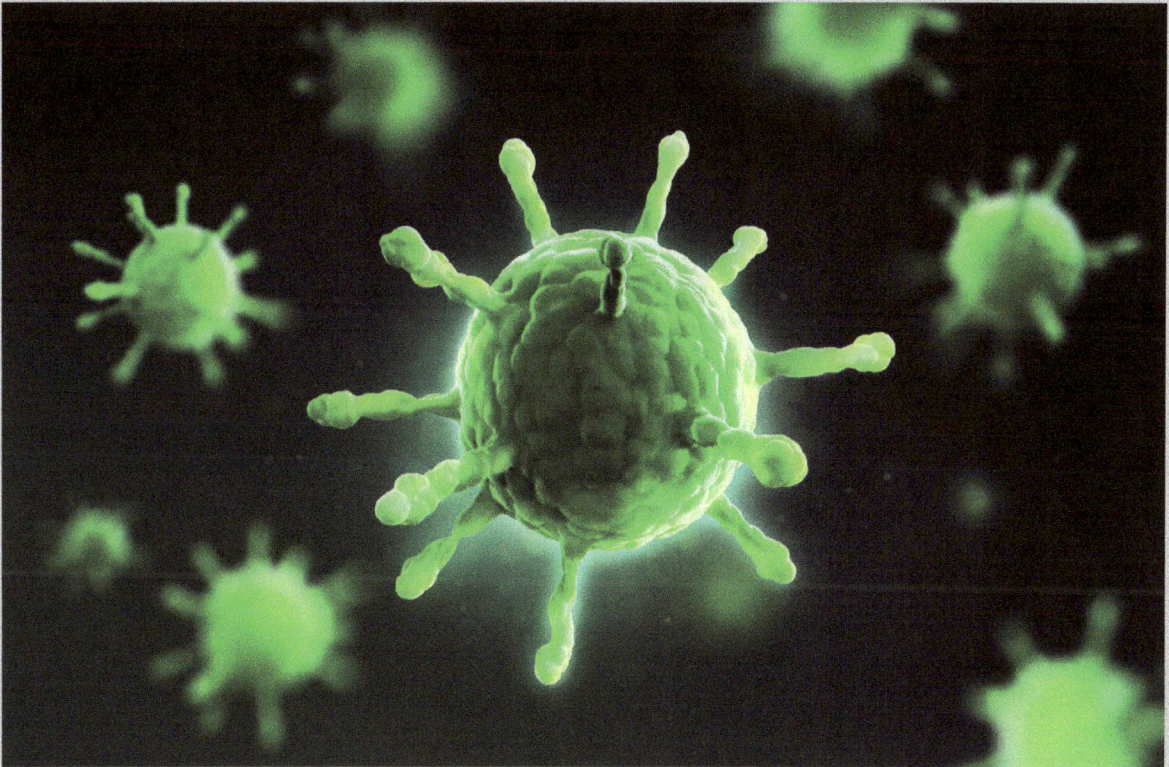

current

the flow of electricity

data

the observations and measurements from an experiment

erosion

the wearing away of rock or soil from the flow of water

friction

the rubbing of one object against another object

galaxy

a group of millions of stars

habitat

the natural home for a plant or animal

invertebrate

an animal without a backbone

joule

the measurement for energy

kinetic energy

the type of energy in moving things

lichen

a patch of fungus that grows on the bark of a tree

mixture

the combination of two or more substances

nectar

the juice inside a flower that a bee uses to make honey

organism

any living thing like plants and animals

predator

an animal that hunts and kills other animals for their food

quantum physics

the science that explains how everything works

reptiles

an animal that lays soft eggs on land and has scaly skin

solar system

a star and all of the planets that orbit or circle around it

temperature

the measurement of how hot or how cold something is

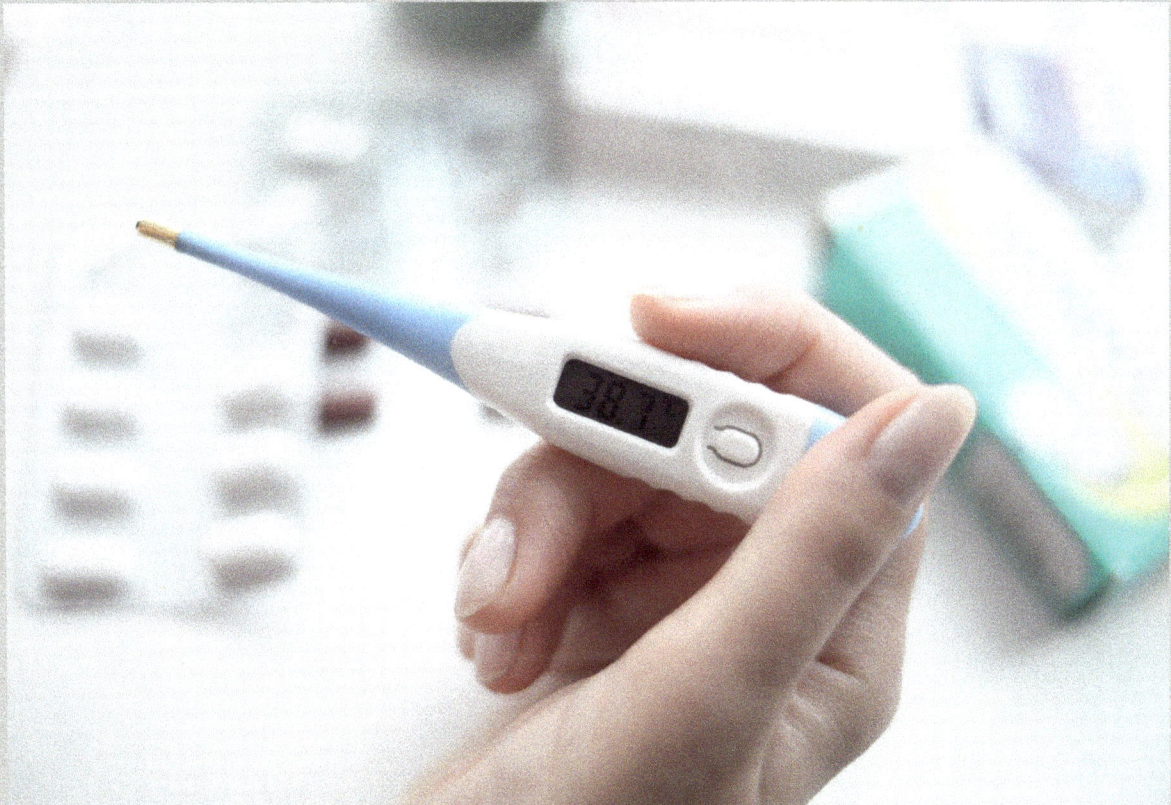

umbra

the shadow on the earth or the moon during an eclipse

variables

the things that change during an experiment

waterproof

something that prevents water from passing through

x-ray

an invisible form of light that allows us to take pictures of broken bones

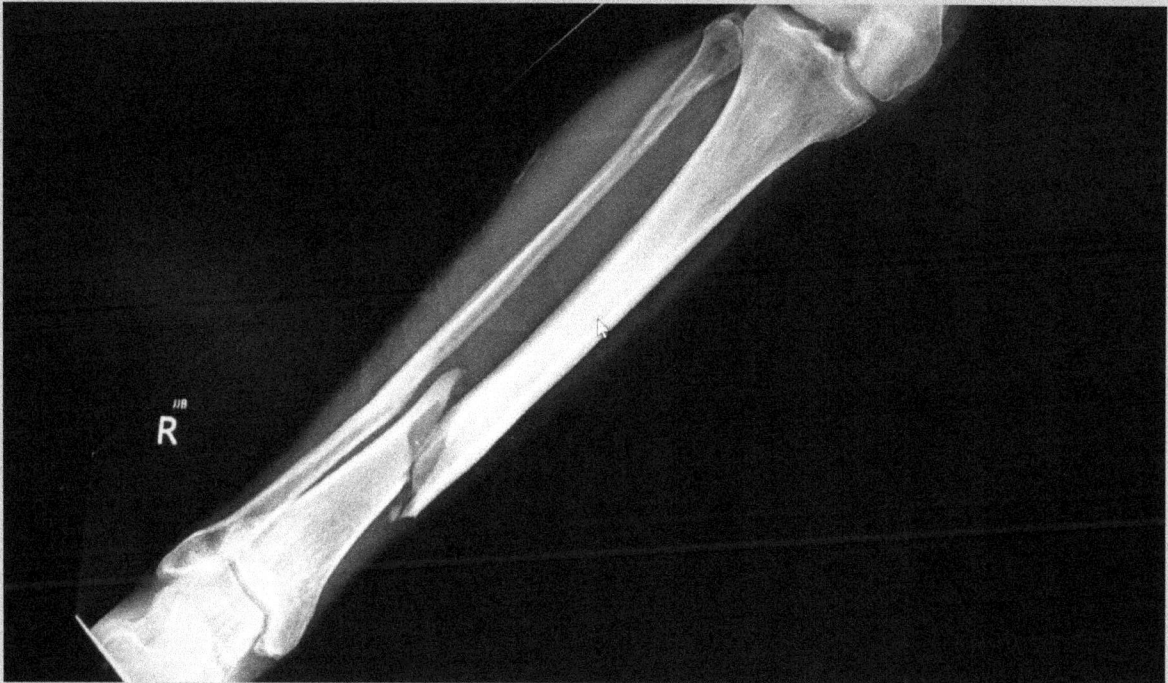

yotta

the weight of the earth in kilograms

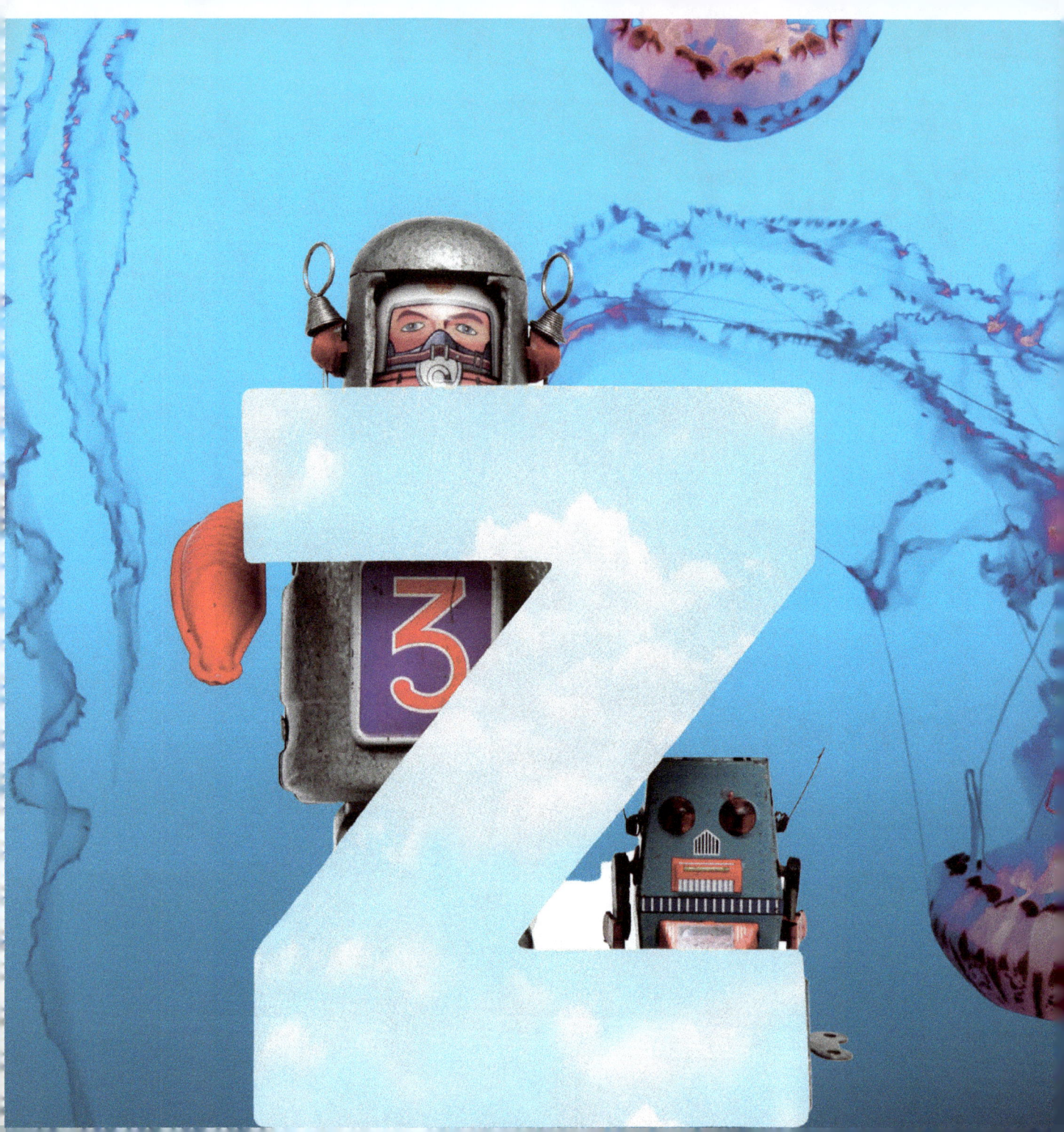

zoology

the study of different types of animals

About the Author

Liz Palmieri-Coonley is an author, 200-hour Registered Yoga Teacher, and Integrative Nutrition Health Coach. She is also a Certified Kids Yoga Teacher with Kidding Around Yoga.

She grew up with a fascination for the human body, illness, and healing, which may seem weird for a kid, but it makes total sense considering both of her parents were in the medical field.

In addition to her love for educating others on yoga and holistic health, Liz is passionate about traveling, spending time outdoors, and everything about her incredible nieces. Some of her more epic adventures include: scuba diving, climbing a waterfall, and riding in a hot air balloon.

Your 1st Experiment

- Blow up a balloon
- Rub the balloon on your shirt for a few minutes
- Bring the balloon up to your hair and see what happens

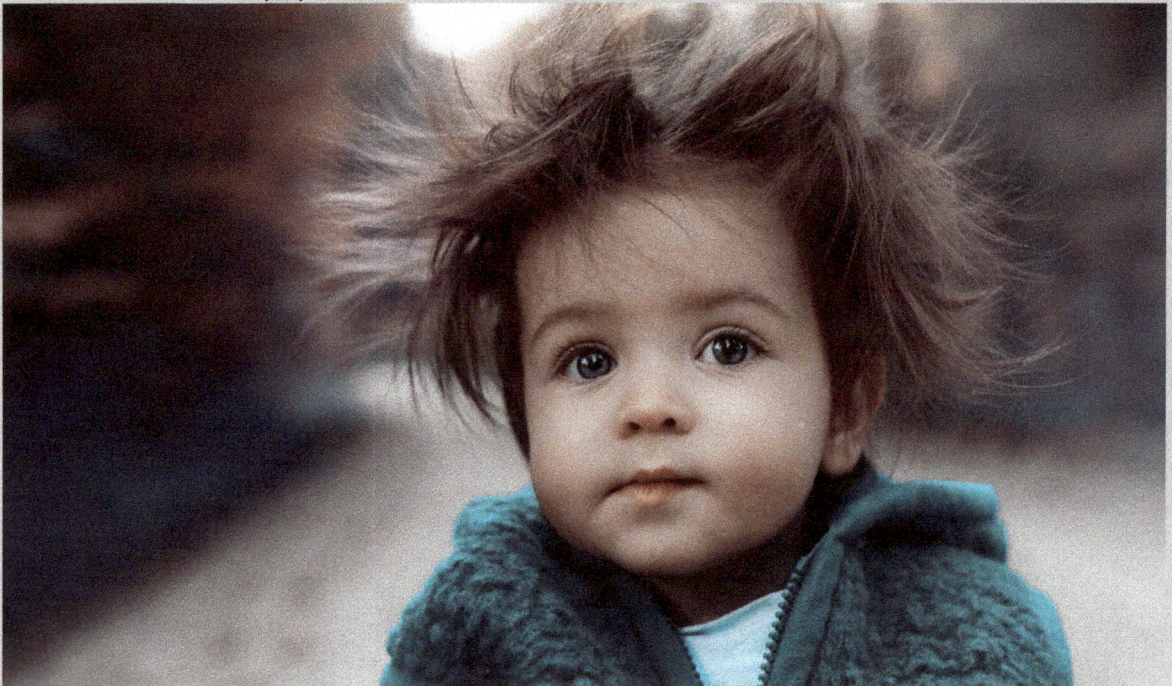

www.ingramcontent.com/pod-product-compliance
Lightning Source LLC
Chambersburg PA
CBHW052339210326
41597CB00031B/5307